THIS BOOK BELONGS TO

FOR MORE INFORMATION, PLEASE CONTACT FUNPORIUM AT
CONTACT@FUNMANDI.COM

THIS BOOK IS DEDICATED TO OUR 4 YEAR OLD "RUKU" WHO MAKES US SO PROUD .

- MOMMY AND DADDY

Count and Trace

1

Count and Trace

2

Count and Trace

3

Count and Trace

4

4 4 4 4 4

4 4 4 4 4

4 4 4 4 4

Count and Trace

5

Count and Trace

6

Count and Trace

7

Count and Trace

8

Count and Trace

9

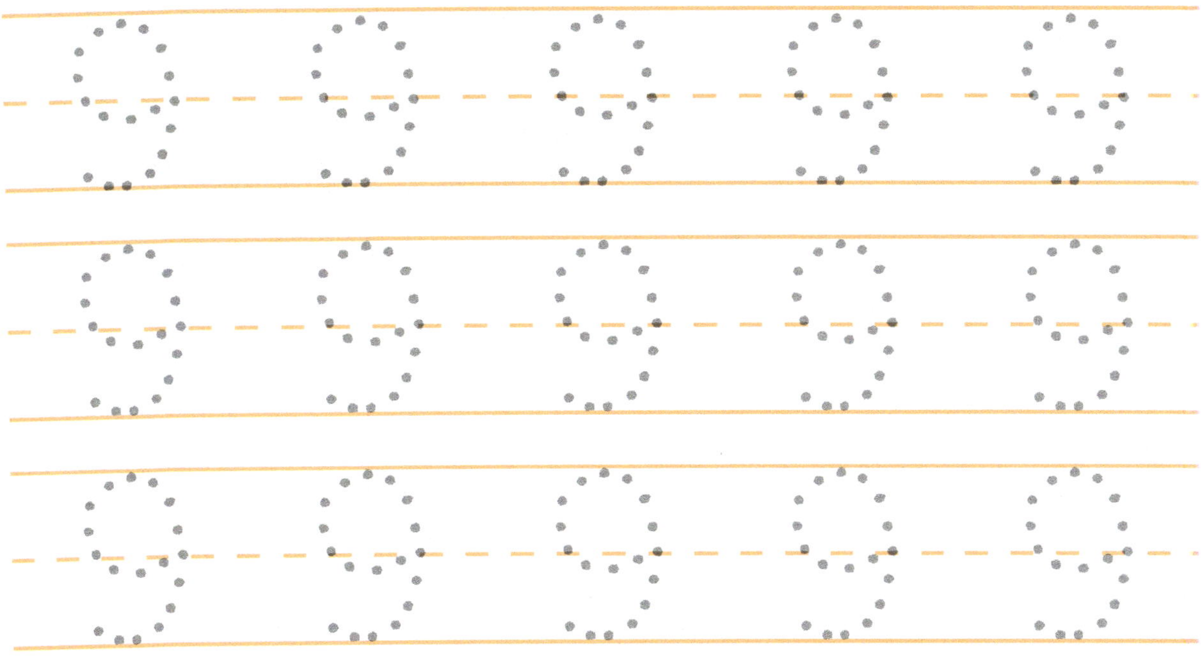

Count and Trace

10

10 10 10 10

10 10 10 10

10 10 10 10

Trace the Numbers

Directions: Count the vehicles and trace the numbers.

	1 1 1 1 1 1
	2 2 2 2 2
	3 3 3 3 3
	4 4 4 4 4
	5 5 5 5 5

Count and Match

Directions : Count the objects and match with the correct number.

1

2

3

4

5

Count and Match

Directions : Count the objects and match with the correct number.

6

7

8

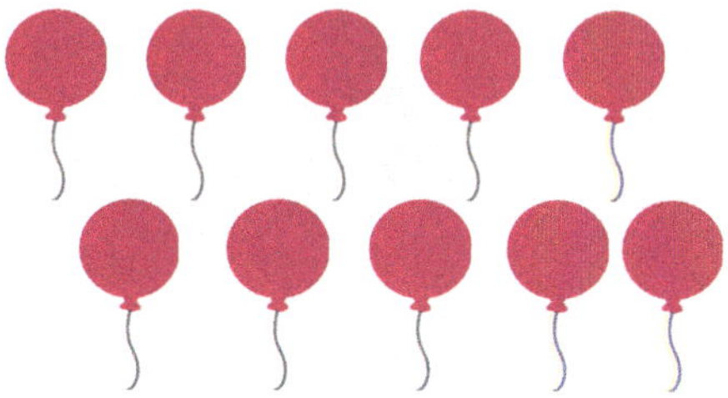

9

10

Count and Match

Directions :Draw a line to match the number with the correct number of flowers.

5

2

6

3

8

4

Count and Match

Directions :Draw a line to match the number with the correct number of dinosaurs.

5

6

4

2

3

1

Count and Match

Directions : Count the insects and match it with the correct number in the box.

1

3

2

Count and Match

Directions : Count the animals and match it with the correct number in the box.

4

5

6

Count and Match

Directions : Count the fruits and match it with the correct number in the box.

8

9

7

10

Count the Vehicles

Directions : Count the vehicles and match with the correct number.

4 3

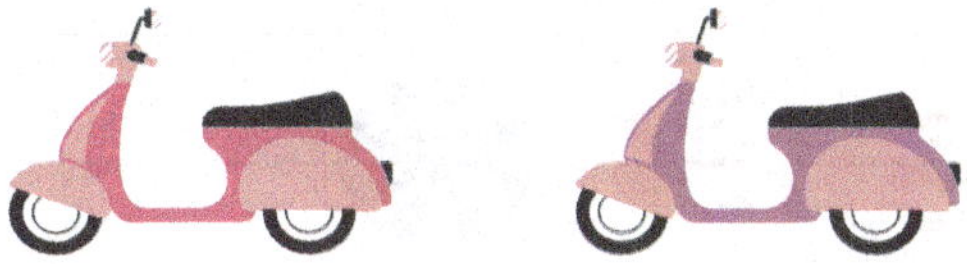

2 1

6 4

5 7

10 8

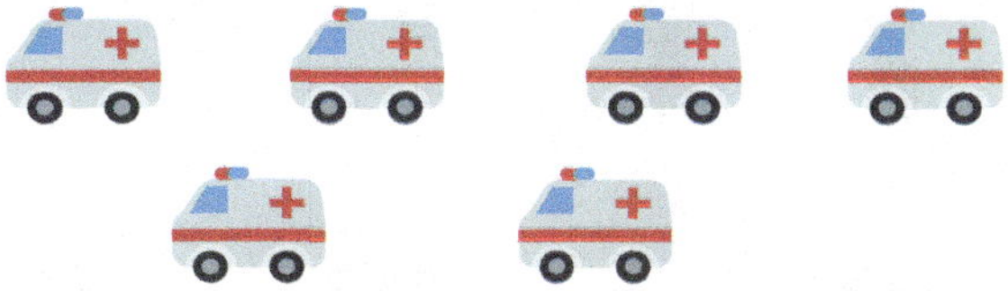

9 6

7 5

3 1

How Many?

Directions : Count the candies and write the correct number in the box.

How Many?

Directions : Count the cool treats and write the correct number in the box.

How Many?

Directions : Count the farm animals below and write the correct number in the circle.

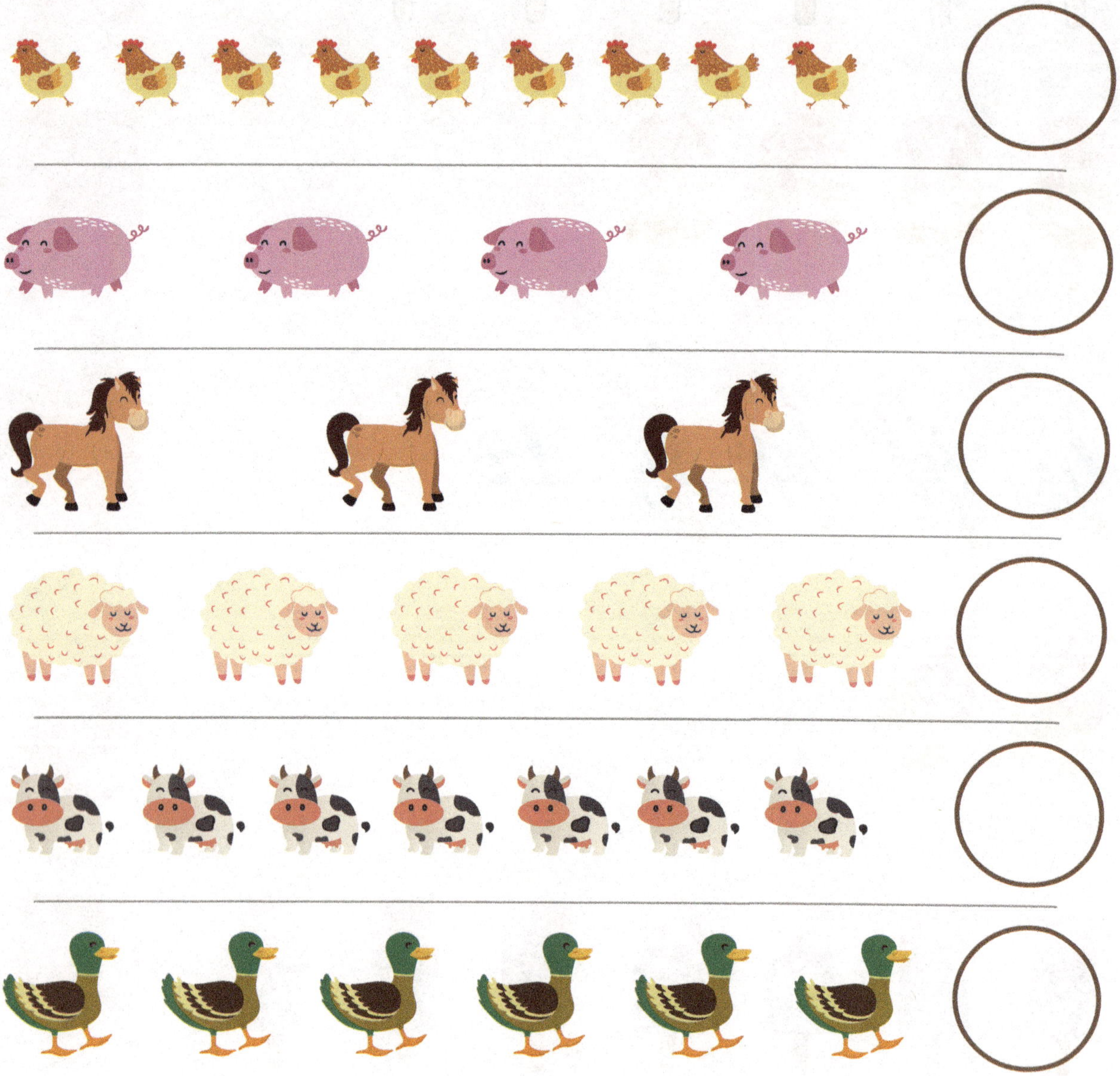

How Many ?

Directions : Count the snacks below and write the correct number in the box.

Ball Addition

Directions : Count and add the balls, then write the answer in the box.

 +  =

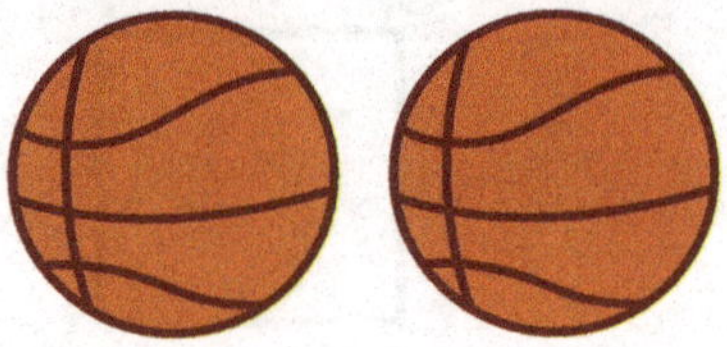 +  =

 +  =

 +  =

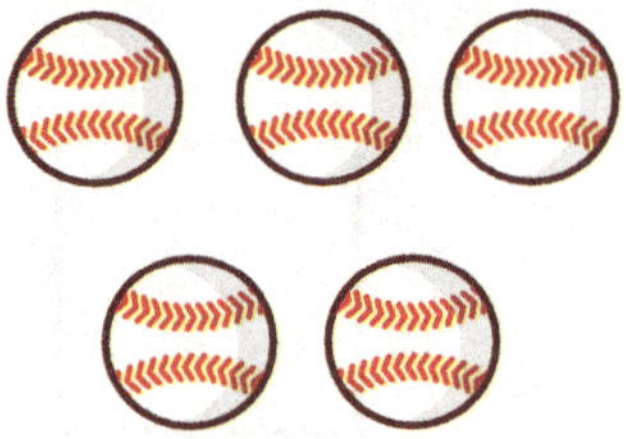 +  =

More or Less

Directions : Circle the group that has more number of objects.

More or Less

Directions : Circle the group that has more number of objects.

More or Less

Directions : Circle the group that has less number of objects.

More or Less

Directions : Circle the group that has less number of objects.

After Numbers

Directions : Help the rabbit to hop 1 step forward.

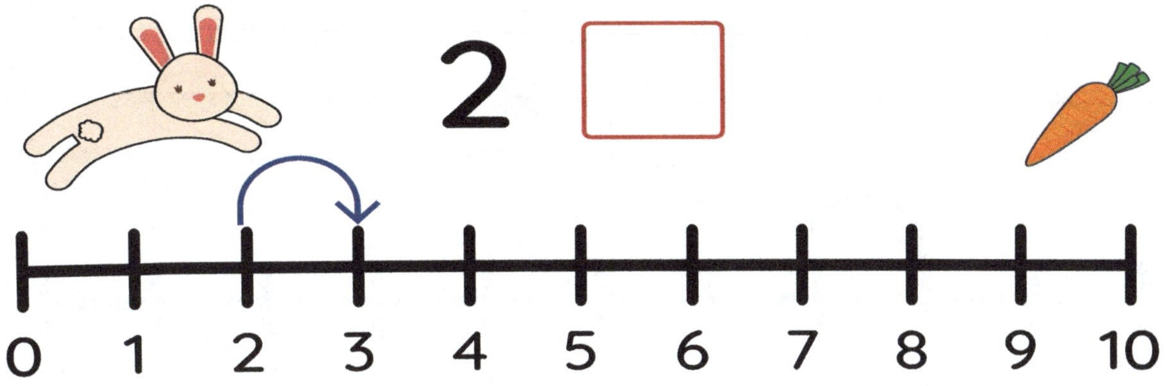

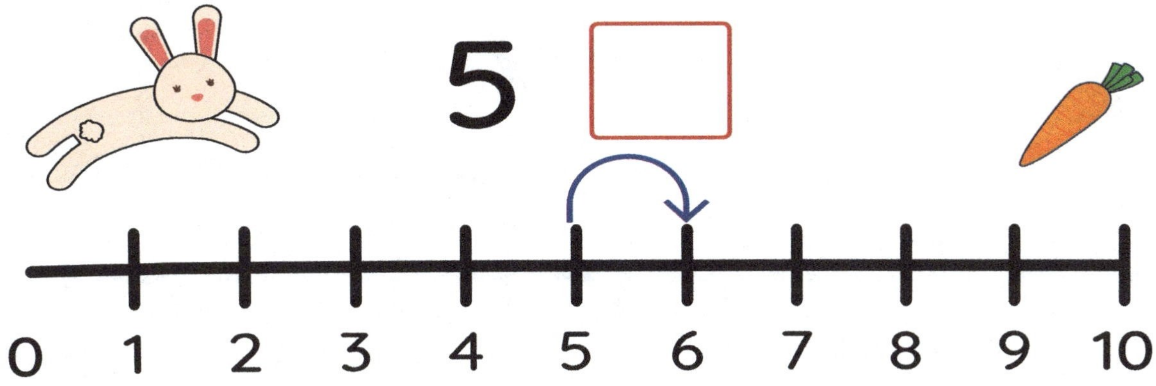

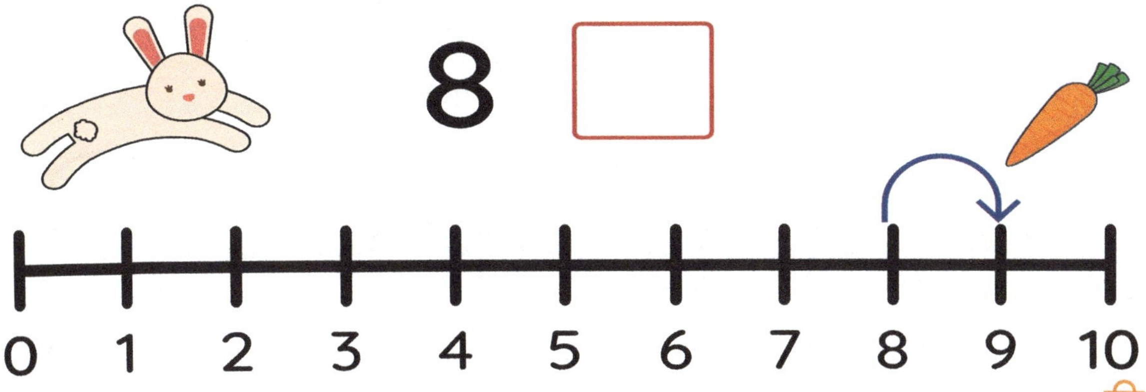

Before Numbers

Directions : Help the rabbit to hop 1 step backward.

☐ 3

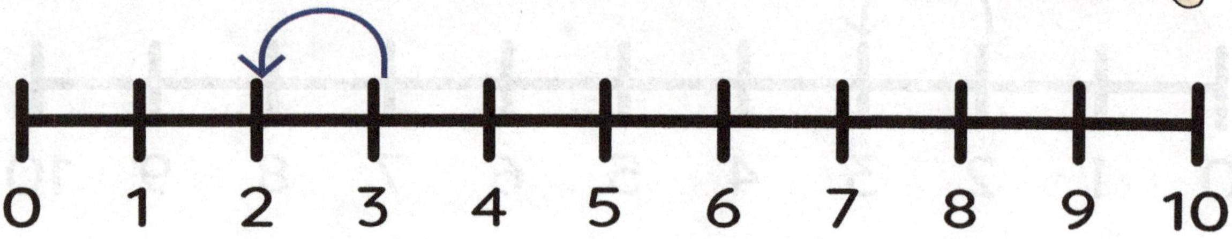

☐ 7

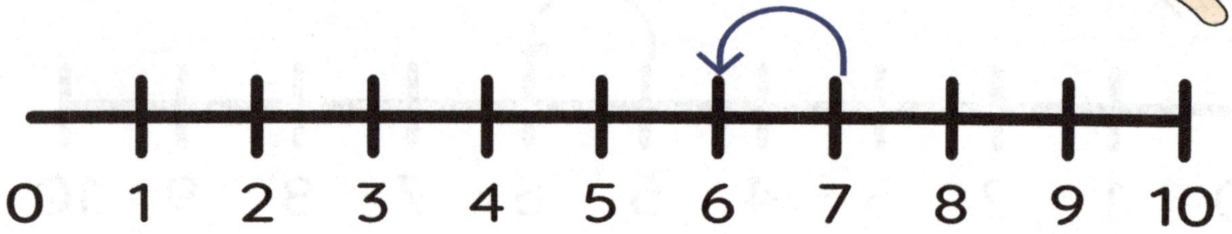

☐ 10

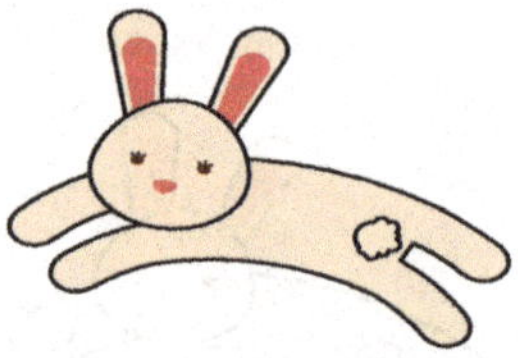

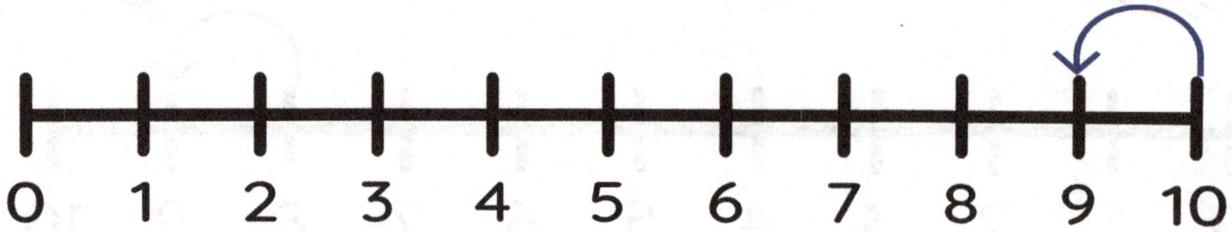

What Comes After

Directions : Use the number line to say what will come after each number.

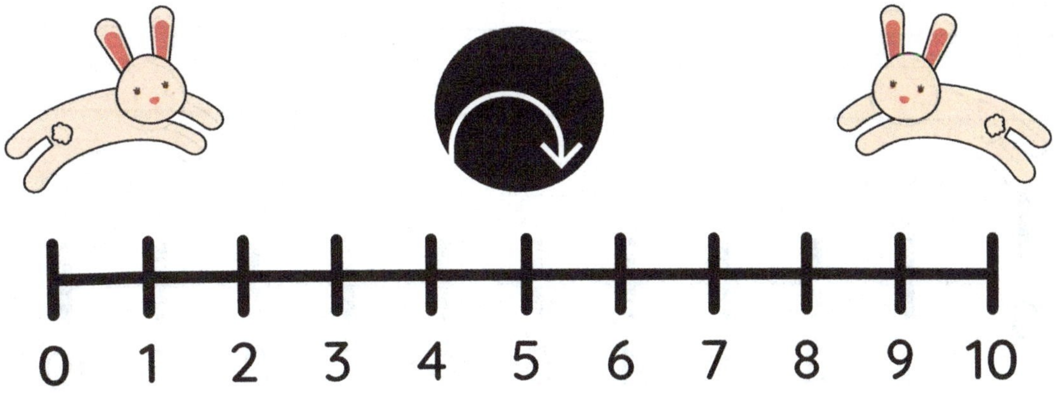

3	
8	
5	

9	
1	
9	

What Comes Before

Directions : Use the number line to say what will come before each number.

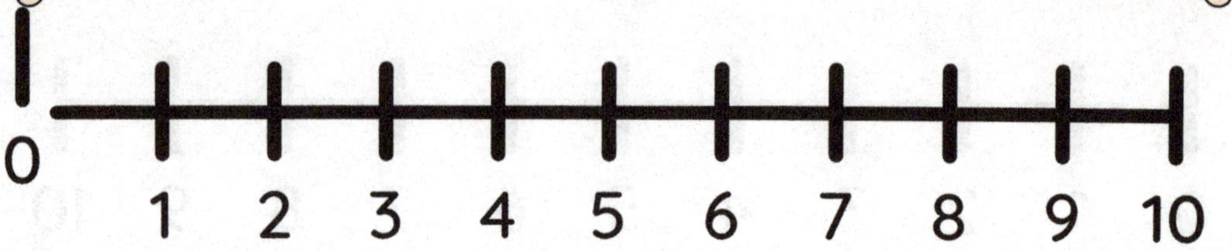

2

6

10

8

5

3

Hat Addition

Directions : Count and add the hat, then write the answer in the box.

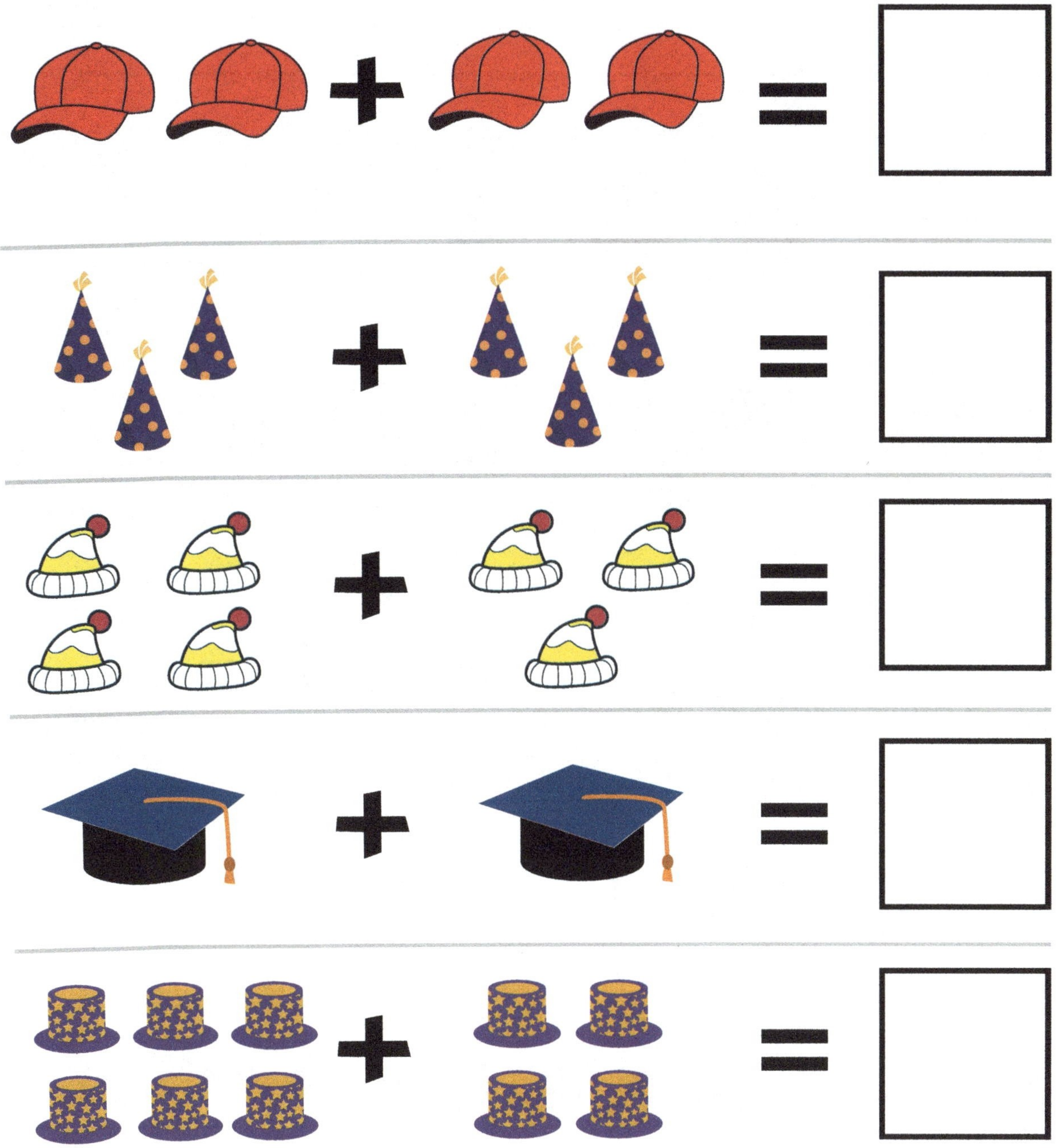

Fruit Subtraction

Directions : Cross out the correct number of fruits to solve each subtraction.

3 - 1 = ______

2 - 1 = ______

4 - 2 = ______

5 - 2 = ______

6 - 3 = ______

Vegetable Subtraction

Directions : Cross out the correct number of vegetables to solve each subtraction.

6 - 3 = ______

5 - 3 = ______

8 - 4 = ______

7 - 2 = ______

Adding Shapes

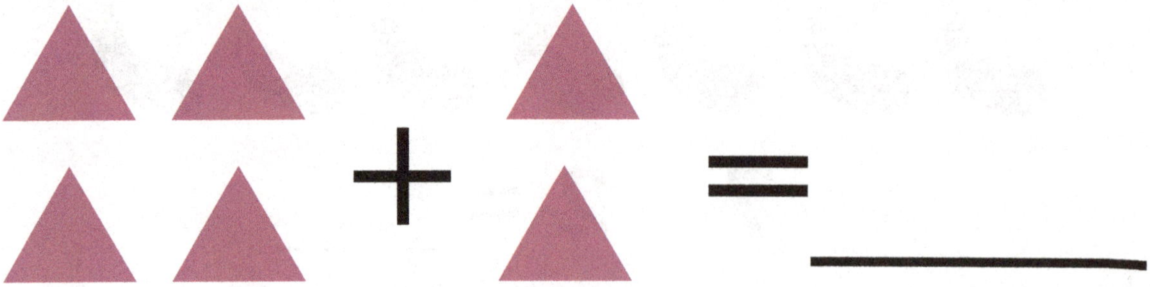

Subtracting Shapes

 − 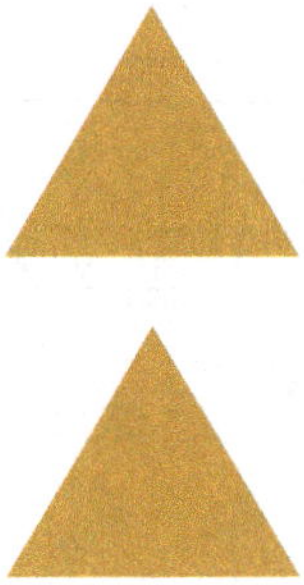 = ______

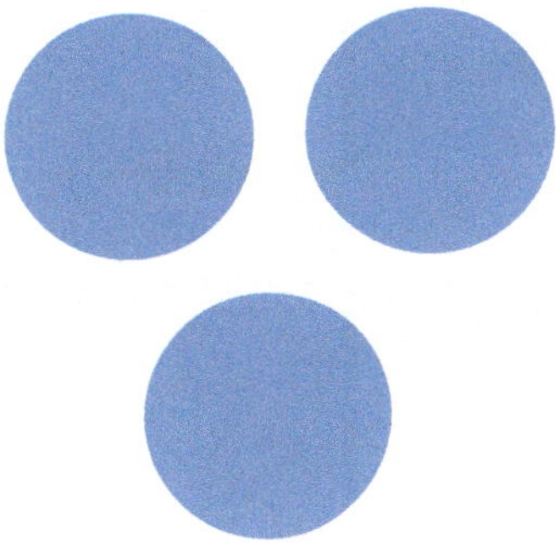 − 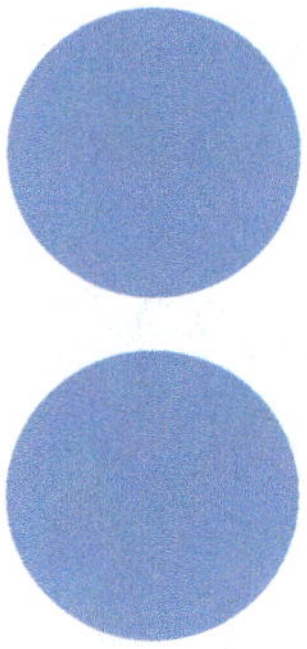 = ______

 −  = ______

Finger Addition

Directions : Add the numbers shown on the hands and write the answer in the space provided.

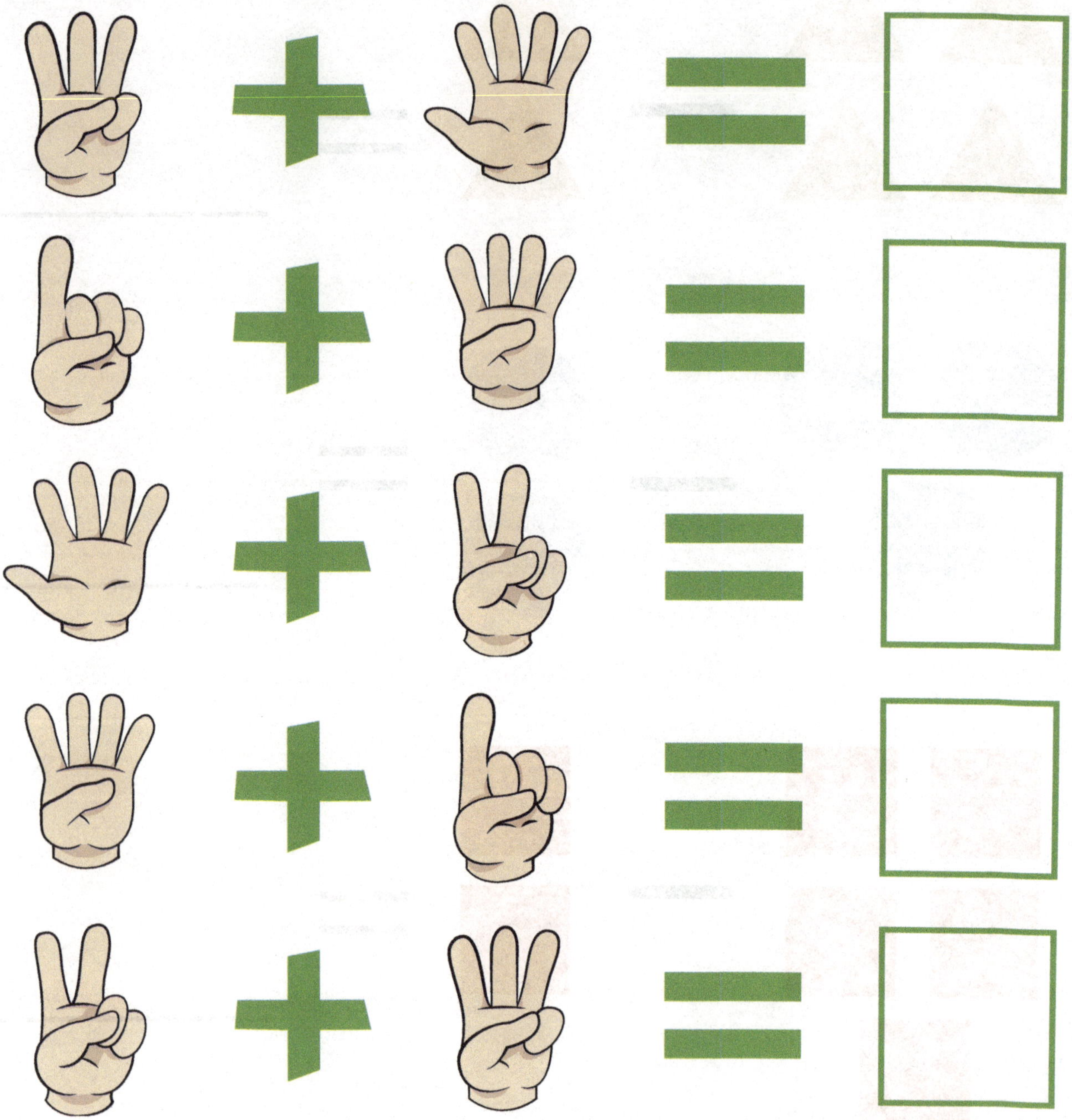

Finger Subtraction

Directions : Subtract the numbers shown on the hands and write the answer in the space provided.

Subtracting Circles

Directions : Cross out the correct number of circles to solve each subtraction.

5 - 2 = ______	◯◯◯◯◯
3 - 2 = ______	◯◯◯
4 - 1 = ______	◯◯◯◯
3 - 1 = ______	◯◯◯
5 - 1 = ______	◯◯◯◯◯
4 - 2 = ______	◯◯◯◯

Odd One Out

Directions : Circle the odd one out.

Odd One Out

Directions : Circle the odd one out.

Odd One Out

Directions : Cross out the one that is different in each row.

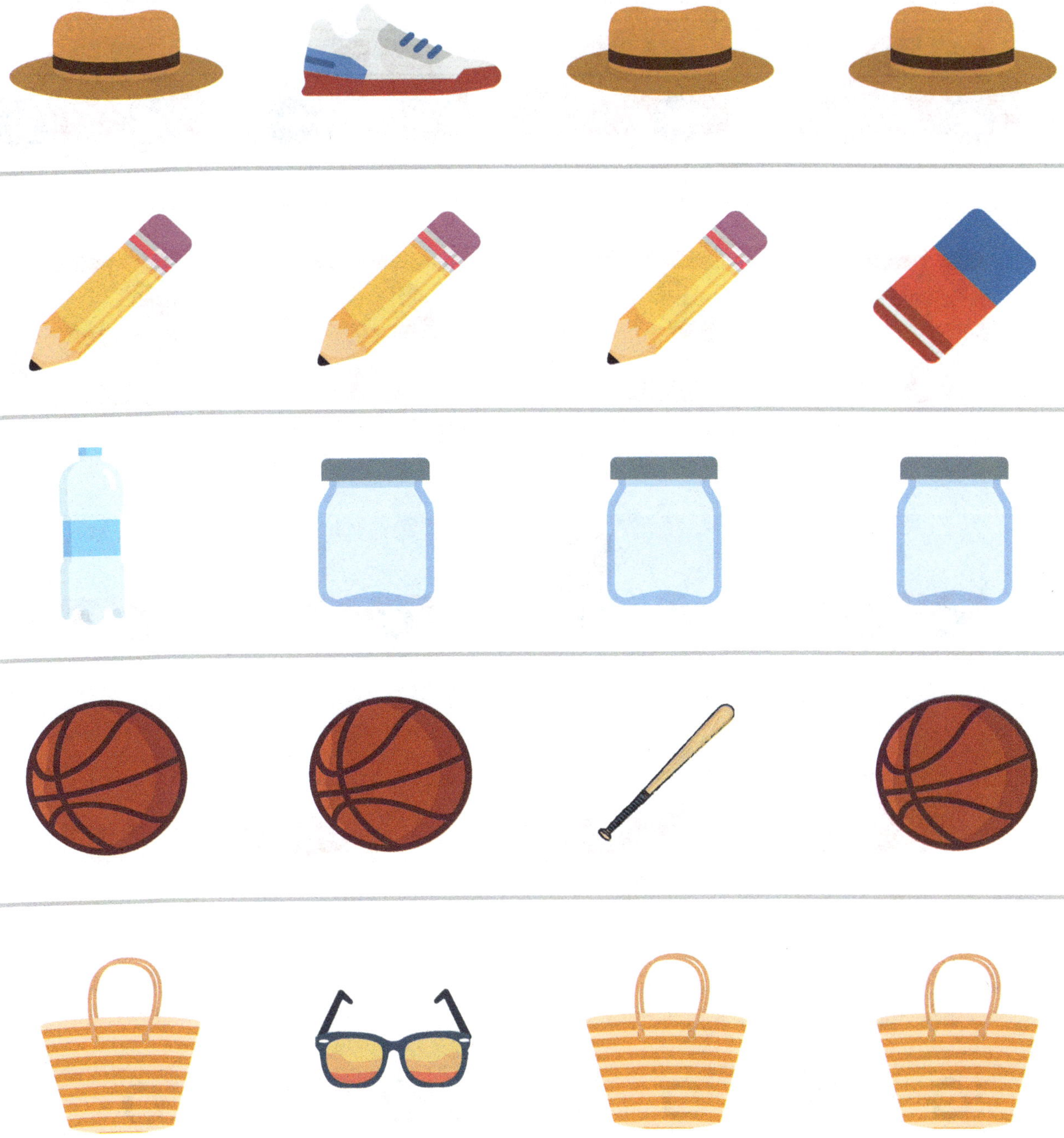

Odd One Out

Directions : Cross out the one that is different in each row.

Same and Different

Directions : Circle the object that is different.

Same and Different

Directions : Circle the object that is different.

What Comes Next ?

Directions : Circle the snack that will come next.

What Comes Next ?

Directions : Circle the shape that will come next.

What Comes Next ?

Directions : Circle the animal that will come next.

What Comes Next ?

Directions : Circle the object that will come next.

What Comes After

Directions : Look at the picture key . Draw what will come after each picture.

What Comes Before

Directions : Look at the picture key . Draw what will come before each picture.

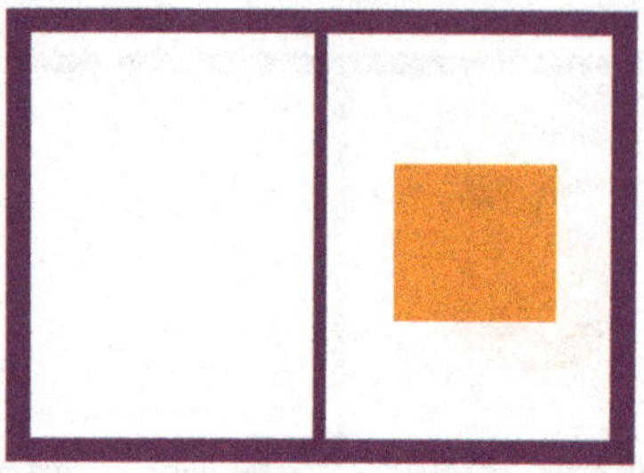

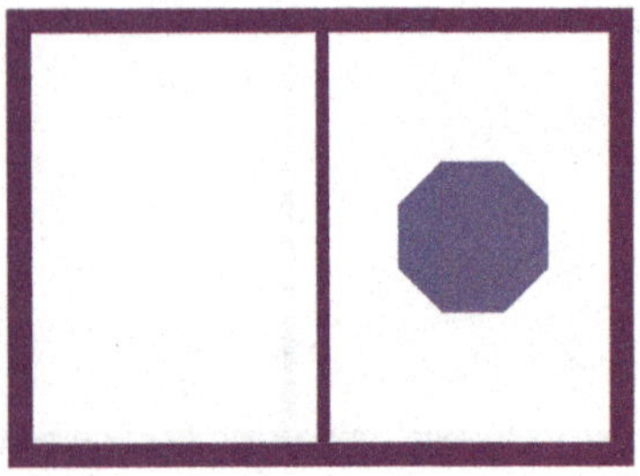

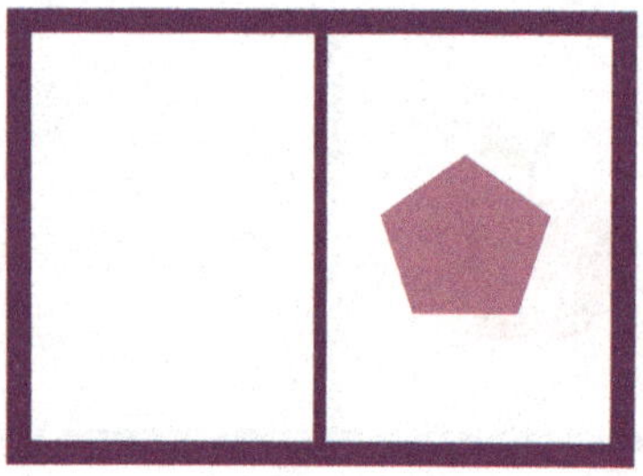

What Comes Next ?

Directions : Circle the Key that comes next.

Sequencing

Directions : Match the correct piece that comes next.

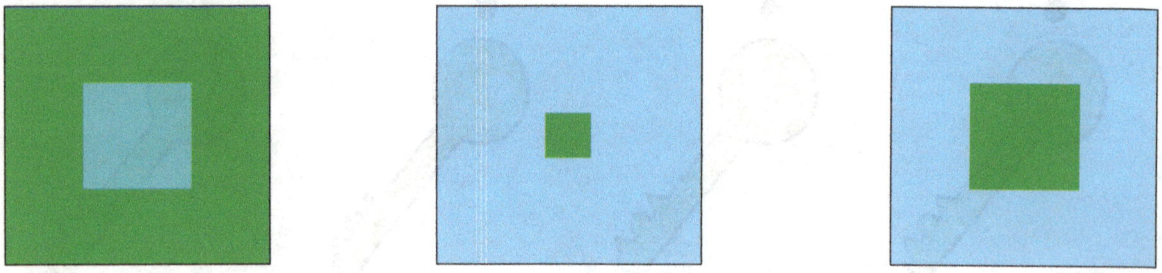

Sequencing

Directions : Draw and color the shape that comes next.

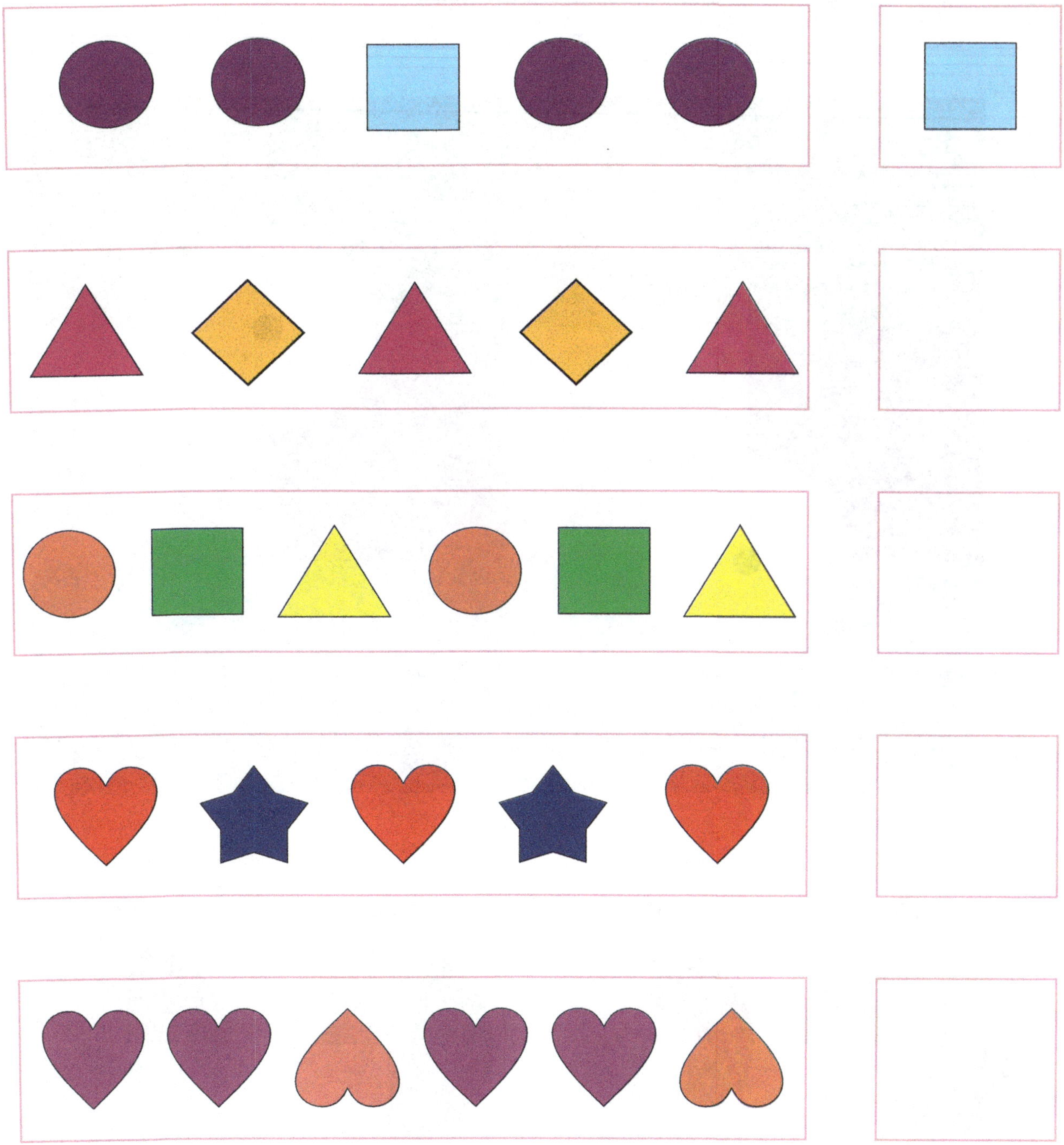

Pattern Completion

Directions : Complete the pattern by matching the correct piece.

Pattern Completion

Directions : Complete the pattern by matching the correct piece.

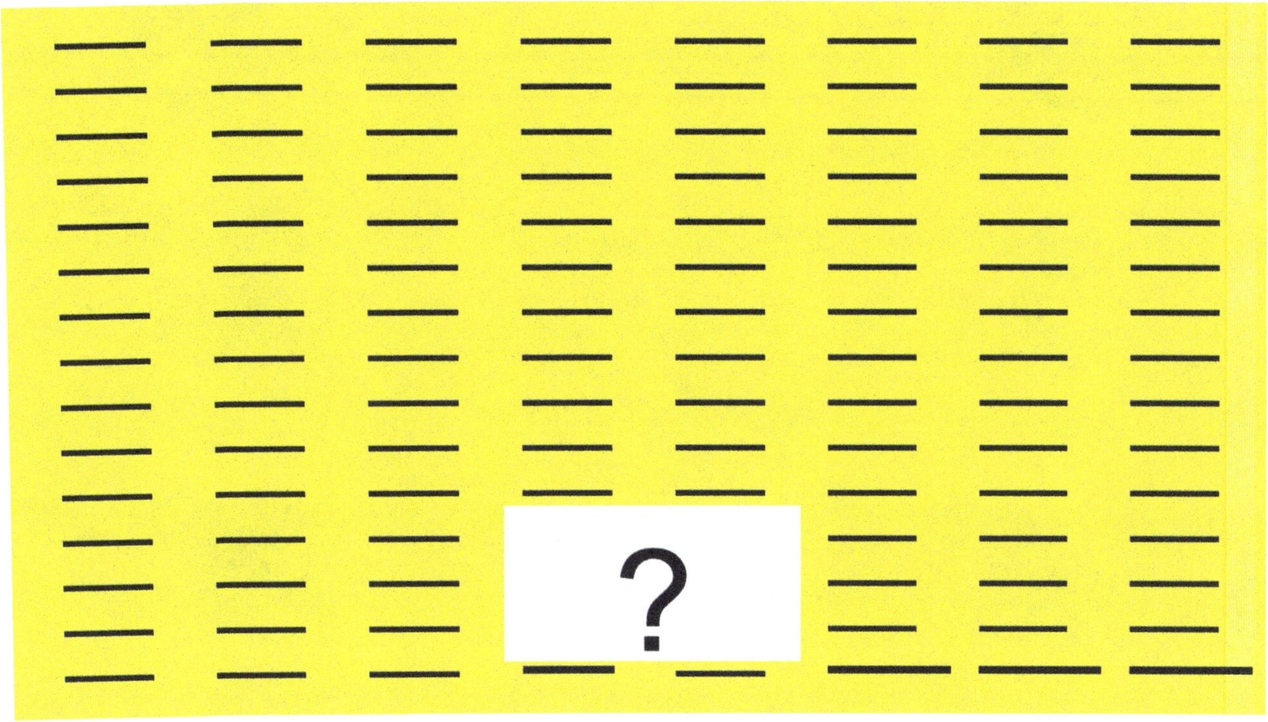

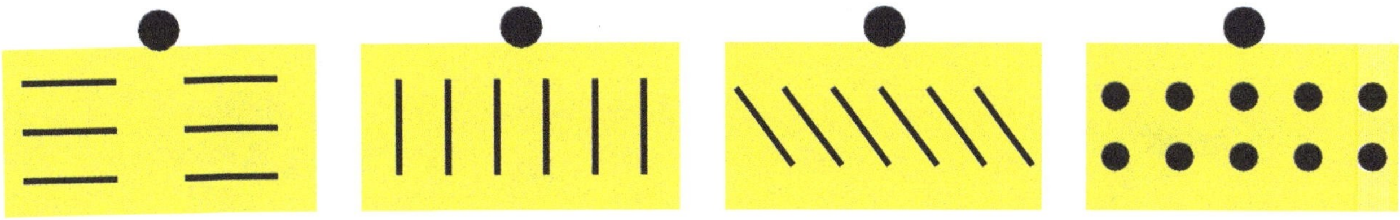

Matching

Directions : Match each mouse to the correct shadow.

Find and Circle

Directions : Pick the shapes to match the picture.

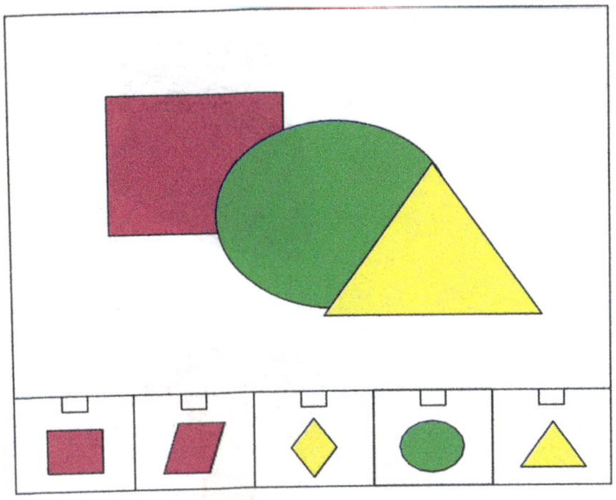

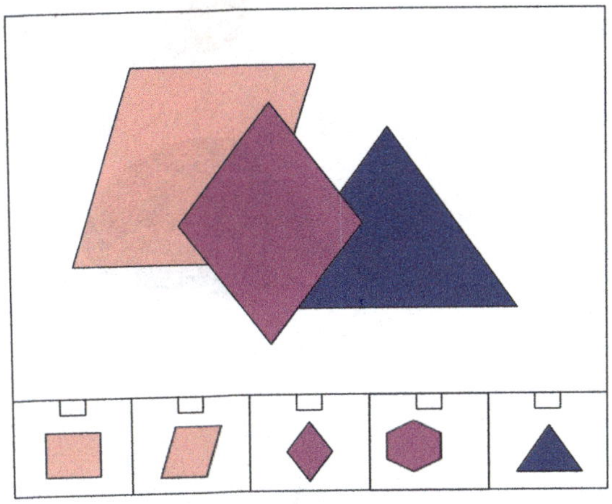

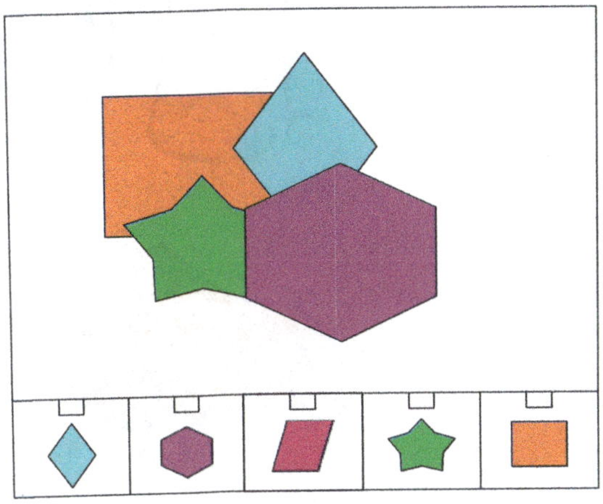

Matching

Directions : Draw lines to match the animal to the correct paw.

Positioning

Directions : Cut out and paste the animals in the correct position.

Ordered Pair

Directions : Write the ordered pair for each item.

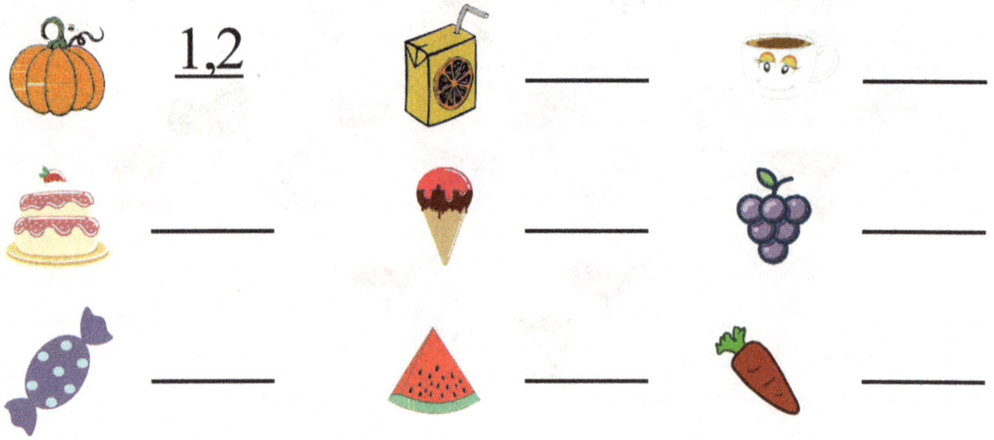

Brain Train

Directions : Cut the objects below and paste on the next page at correct place.

Brain Train

Brain Train

Directions : Additional Practice sheets for Numbers 1-10

Brain Train

Directions : Additional Practice sheets for Numbers 1-10

Brain Train

Directions : Additional Practice sheets for Numbers 1-10

Brain Train

Directions : Additional Practice sheets for Numbers 1-10

Brain Train

Directions : Additional Practice sheets for Numbers 1-10

Brain Train

Directions : Additional Practice sheets for Numbers 1-10

Brain Train

Directions : Additional Practice sheets for Numbers 1-10

Illustrations:
Independent artists
Stock images @ Vecteezy.com and canva.com

Made in the USA
Monee, IL
02 November 2026

67375619R00044